SPACE WONDERS
GALAXIES
I0815773
BLACK RABBIT BOOKS

TABLE OF CONTENTS

1

Milky Way

Look up at the night sky! See that dim, white band of light? That is the Milky Way. It is our galaxy. A galaxy is a big group of stars, gas, and dust. The Milky Way is shaped like a spiral. Two arms swirl around its center.

It is hard to see all of the Milky Way. Scientists use special **telescopes**. These take pictures from space.

Think About It

Many galaxies are spirals.
Why do you think this is true?

2 Andromeda

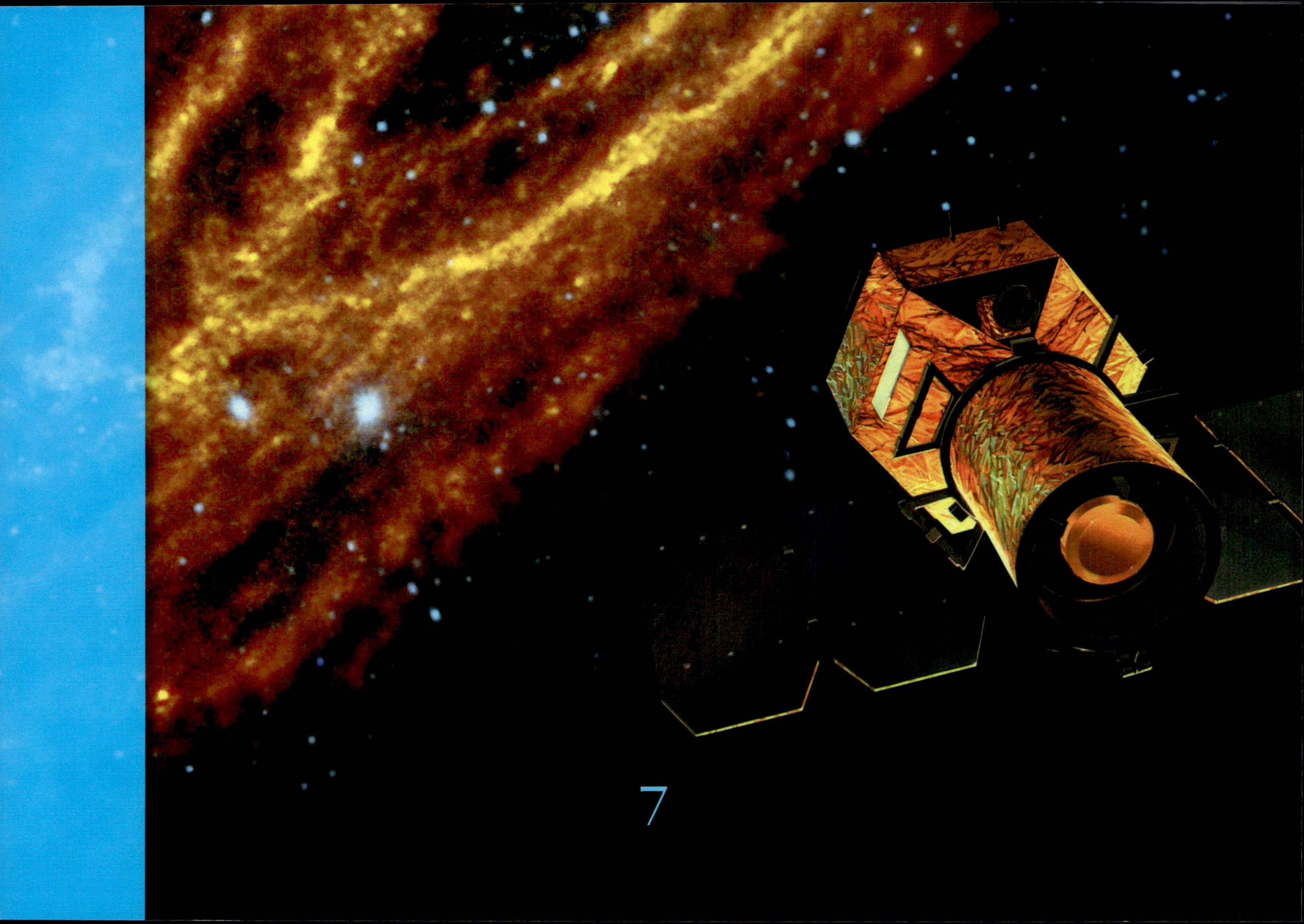

Andromeda is our neighbor. It is the nearest galaxy to the Milky Way. It is about 2.5 million **light-years** from Earth. It might have formed when two smaller galaxies crashed together.

Look at the sky in August or September. Face north. Find a "W" shape made of five stars. That is a **constellation**. Its name is Cassiopeia. It points to the Andromeda galaxy.

3

Bode's Galaxy

Bode's Galaxy is part of the Great Bear star group. It is 11.6 million light-years from Earth. It is a spiral. It has a huge **black hole** at its center. Like other galaxies, its arms are full of stars. These stars are millions of years old.

The *Hubble* telescope took photos of this galaxy. It took two years to capture it all! The stars look like tiny grains of sand. Some stars may still be forming.

Hubble Space Telescope

Think About It

How can photos help us learn about galaxies?

Sombrero

4

This galaxy is named for its shape. It has a round center. Its edges are flat. It looks like a sombrero. That's a Mexican hat.

Dust spins around the galaxy's center. It creates dark edges. Scientists think a black hole is at its center. It would be the biggest black hole ever found in a galaxy.

Did You Know?
This galaxy was discovered in 1781.

5

Pinwheel

It is easy to see how this galaxy got its name. It looks like a pinwheel toy. It is one of the closest galaxies to the Milky Way. It is almost twice as big!

Pinwheel's arms have **nebulas**. These are giant clouds. They are full of **hydrogen**. This gas creates new stars. Pinwheel has more than 3,000 of these clouds. The new stars burn hot and bright.

Did You Know?
There is no black hole in Pinwheel's center.

Leo Triplet

Why have one galaxy when you can have three? Leo Triplet is a group of three galaxies. They are in the Leo constellation. They are very close together. Gravity connects them. They tug and pull at each other.

Each galaxy is tipped at a different angle. They look like they have different shapes. But they are all spirals.

This group is very far away. It is about 35 million light-years from Earth.

Think About It

How does our position on Earth affect what we see in space?

MORE TO EXPLORE

FANTASTIC FACTS

It takes 250 million years for our sun to move around the Milky Way one time.

Andromeda's center has a ring of old, red stars and young, blue stars.

Hubble took a photo of the Pinwheel Galaxy. It is the biggest photo ever taken of a galaxy.

Photos of the Sombrero galaxy were used in a 1960s TV show called *The Outer Limits*.

Bode's Galaxy is also known as Messier 81.

One of the Leo Triplets is called the Hamburger Galaxy because of its shape.

MORE TO EXPLORE

COOL COMPARISONS

Which galaxy is the biggest across?

MORE TO EXPLORE

RESOURCES

Glossary

black hole (BLAK HOHL) An invisible area in space with gravity so strong that light cannot get out of it.

constellation (kon-stuh-LEY-shuhn) A named group of stars that forms a shape in the sky.

hydrogen (hahy-druh-juhn) A chemical element that has no color or smell.

light-year (LAHYT-yeer) A distance equal to the distance that light travels in one year; one light-year is about 5.9 trillion miles (9.5 trillion kilometers).

nebula (NEB-yuh-luh) A group of stars that look like a bright cloud at night.

telescope (TEL-uh-skohp) A tube-shaped instrument that you look through to see things that are far away.

Read More

Corrigan, Eamonn. *The Milky Way and Other Galaxies.* Buffalo, NY: Norwood House Press, 2025.

Dittmer, Lori. *Galaxies.* Mankato, MN: Creative Education and Creative Paperbacks, 2025.

Index

TOP RANK is published by Black Rabbit Books, P.O. Box 227, Mankato, MN, 56002. • Top Rank is an imprint of Black Rabbit Books • Series designed by Danny Nanos • Book designed by Jason Knudson • Photographs © Adobe Stock/Anjali, 8–9; Freepik/EyeEm, 17, rawpixel.com, 2–3; NASA, 23; NASA/ESA/Herschel/PACS/SPIRE/J. Fritz, U. Gent; X-ray: ESA/XMM Newton/EPIC/W. Pietsch, MPE, 6–7, 23; NASA/JPL-Caltech, cover; NASA/JPL-Caltech/ESA/STScI/CXC, 16–17, 23; Science Source/Detlev Van Ravenswaay, 14–15; Shutterstock/MarcelClemens, 13, Vasilyev Alexandr, 8; Wikimedia Commons/Benjamin Inouye, 4–5, ESO/Yuri Beletsky, 20–21, Fried Lauterbach, 19, Giuseppe Donatiello, 10–11, 23, Kurtzepp, 20–21, 23, NASA, 7, 11, 12, NASA, ESA, and the Hubble Heritage Team (STScI/AURA), 18, NASA/JPL-Caltech/R. Hurt (SSC/Caltech), 2, 23, NOIRLab/NSF/AURA, 12 • Printed in the United States of America

Library of Congress Cataloging-in-Publication Data: Names: Mattern, Joanne, 1963- author. | Title: Galaxies / Joanne Mattern. | Description: Mankato, MN: Black Rabbit Books, [2026] | Series: Space wonders | Includes bibliographical references and index. | Audience: Ages 8–11 | Audience: Grades 2–3 | Identifiers: LCCN 2025011960 (print) | LCCN 2025011961 (ebook) | ISBN 9781645825098 (library binding) | ISBN 9781645825272 (paperback) | ISBN 9781645825456 (ebook) | Subjects: LCSH: Galaxies—Juvenile literature. | Classification: LCC QB857.3 .M28 2026 (print) | LCC QB857.3 (ebook) | DDC 523.1/12—dc23/eng/20250324 | LC record available at https://lccn.loc.gov/2025011960